MODBUS

The Everyman's Guide to a Protocol That Has Stayed Relevant in Automation for Over 30 years

JOHN RINALDI

JOHN RINALDI

ISBN-13: 978-1517764685
ISBN-10: 1517764688

DEDICATION

To the Automation Engineer, the unsung hero of American Manufacturing.

JOHN RINALDI

TABLE OF CONTENTS

TABLE OF FIGURES

TABLE OF TABLES

ACKNOWLEDGMENTS

This book would not be possible without the dedication, friendship, persistence, support and follow through of the entire staff at Real Time Automation. Specifically, to Drew Baryenbruch for pushing me to do this book and freeing me of daily sales and marketing so that I can take on projects like this.

FOREWORD

Modbus Changed the World!

The Modbus communications protocol is the networking granddaddy of the industry and is still the only open source electronic networking protocol for connecting automation systems. Modbus has stood the test of time and is still being used in a wide range of applications, including industrial automation, process control, building automation, transportation, energy, and remote monitoring. Virtually any type of sensor and controller devices can be found that incorporate Modbus networking, including programmable logic controllers (PLCs), process controllers, process instruments, process sensors, PID controllers, motor drives, energy meters, Supervisory Control and Data Acquisition (SCADA) systems, programmable automation controllers (PACs), discrete sensors, valves, and many other embedded devices.

The key to the success of Modbus includes simplicity, ease of implementation, and open source for anyone to use. Modbus is used throughout the world, and understanding how to apply it as described in this book is important for any person involved in the automation industry.

THE EVERYMAN'S GUIDE TO MODBUS

The Modbus serial communication protocol was developed by Modicon and published by the company in 1979 for use with its programmable logic controllers (PLCs). The early roots of Modicon started in 1968 with a core group of engineers led by Dick Morley that invented the first programmable logic controller.

In simple terms, Modbus is a method used for transmitting information over serial lines between electronic devices. Originally intended for communications between programmable logic controllers (PLCs) and computers, it has become a de facto standard communication protocol for connecting a wide range of industrial electronic devices.

Modbus is an extremely compact and flexible protocol that continues to prove it can be adapted for use in a wide range of applications and media.

Modbus is popular for remote applications that communicate over almost any means, including wired and cellular telephone, licensed and unlicensed radios, and satellite. Developers have used Modbus to leverage other wireless communications technologies, such as in the Industrial, Scientific and Medical (ISM) radio band, short message service (SMS), General Packet Radio Service (GPRS), and Mesh networking (802.15.4). Modbus is incorporated into Remote Terminal Devices (RTUs), used in applications including water/waste water, oil & gas, electric power utility, cell tower monitoring, and tank monitoring.

In November 2007, ODVA extended the CIP Networks Library of specifications to provide compatibility of Modbus TCP devices by extending the Common Industrial Protocol (CIP) Network specifications to provide compatibility of Modbus/TCP devices with networks built on the CIP. This extension provides Modbus/TCP users a clear path to integrate with CIP network architectures. Users benefit through interoperability between the EtherNet/IP and Modbus/TCP devices.

Illustrating the pervasiveness and flexibility of Modbus, the Semiconductor Equipment and Materials International (SEMI) global industry association serving the nano- and microelectronic manufacturing industry in response to customer demand implemented a Network Communication Standard and an Object Messaging Protocol using Modbus TCP/IP. This allows sensors conforming to the organizations SEMI Sensor Bus Standard to communicate with each other using Modbus TCP/IP.

It is almost impossible to walk into a manufacturing plant, process plant, oil platform, commercial building, ship, or other area where there is automation, sensors, and controls that do not have devices communication using Modbus. Modbus is not industry specific and is continually being adapted for use in a wide range of applications. Modbus/TCP is a compact protocol that can be used with new low cost processors that incorporate communications being embedded in sensors and other edge devices. Modbus has stood the test of time and will continue to be used for years to come.

William Lydon
Editor, Automation.com
May, 2015

INTRODUCTION

A Book on Modbus?

When I was approached to write a book on Modbus in 2015, I dismissed the idea outright. Writing a book on Modbus in the age of new factory floor platforms like mobile phones and tablets, new technologies like OPC UA, new businesses paradigms from the Internet of Things (IOT) and the ever present march toward integration between the Enterprise and factory floor just seemed silly.

It seemed to be as useful as writing about different models of buggy whips, the rotary telephone, or that new innovation, color television. What is there to say? What hasn't been said about Modbus over the last 40 years?

Modbus is hardly a new technology. Historians can disagree about its actual birth, but it's certainly a product born of the 1970s. Success is such a trite word for how well it's done over those 40 years.

Modbus has found its way into hundreds of thousands—if not millions—of devices. You can find it in everything from valve controllers, to motor drives, to HMIs, to water filtration systems. It would be difficult indeed to name a product category in Industrial or Building

Automation that doesn't use Modbus.

Yet even in the automation world, Modbus isn't just *old* technology. IT'S ANCIENT TECHNOLOGY.

Modbus is like that loveable old uncle that comes over every Thanksgiving. He's retired now, he putters around his garden, he's no longer the handsome debonair man of 40 years ago, but he's there when we need him and that's why we love him.

Prior to Modbus, all we had was electrical signaling. For digital input and output devices like pushbuttons, lights, motors and the like, we wired a signal wire and a return from the controller to the device. For analog input devices like a temperature sensor or output devices like drive speed, we used 0-10 Volts, 4-20ma current loop or RTD. Everything was wired, everything was either a voltage level, a current level, or simply a binary input or output. With thousands of inputs and outputs in a big manufacturing machine, the labor to install all those wires, check that every single one of them was terminated to the right position, and test each one took weeks and sometimes months of effort.

In those days, the concept of data, let alone information, just didn't exist. We had inputs. We had outputs. Even things that cried out for digital control had to be wired as an analog. Drive speed couldn't be specified as 100 RPM. Instead it was specified as a 0 to 10 Volt output with the precision of that output defining how precisely you could control a speed. Luckily, everything else was pretty crude too so we didn't need very precise control a lot of the time.

But Modbus changed all that. Modbus changed everything. Modbus introduced the concept of data on the factory floor. Modbus made it possible to connect an entire group of devices using only two wires on the controller. That alone saved a massive investment in wire, labor and installation time. Instead of miles and miles of wire connecting hundreds of devices, a simple two-wire pair could be daisy-chained from one device to the next to the next. It was revolutionary for its time.

It wasn't just that Modbus was the first serial protocol. Modbus was the right technology at the right time. You have to remember that the first microprocessor wasn't invented until shortly before the birth of Modbus. Do you remember what those microprocessors were like? Simple 8-bit processors with severely limited code space and memory.

I remember those days. In one of my first installations on a packaging machine in Canada we had 128 bytes of RAM and I manually tracked every single one of those 128 8-bit RAM locations. For example, I tracked byte 14 and knew that it was only used in the first section of the program so that I could reuse it for something else later in the program. Try to imagine how revolutionary Modbus was in this kind of automation environment.

And that's why Modbus achieved so much early success. Its raw simplicity made it the preferred implementation for many other protocols. It had a number of distinct advantages:

1. **Modicon Sponsorship** – It didn't hurt Modbus that one of the biggest PLC manufacturers at the time created Modbus and made it open and widely available.

2. **Simple Data Representation** – Modbus has only two basic data types, 16-bit unsigned integer (known as a register) and a single bit (known as a coil).

3. **Simple Request/Response Command Structure** – Modbus has a simple read and write for each of its different data types.

4. **Low Resource Requirements** – Modbus requires very little in the way of processor code space or RAM. This isn't as important today given the

powerful processors and technology available to us, but it was very important in the early years of industrial automation when processors used 8-bit technology and resources like RAM and ROM were extremely expensive and scarce.

5. **Serial Transport Layer** – Modbus uses RS485 serial communications to move bits from one place to another. This means that any processor can implement Modbus without any special hardware. All you need is a simple and inexpensive RS485 driver chip to be in the Modbus business.

6. **Message Checking** – CRC and LRC checking mean that transmission errors are checked to 99% accuracy.

But the other side of the coin is that all this simplicity leads to some distinct disadvantages including:

1. No Device Profiles

2. Limited data types

3. RS485 problems and troubleshooting difficulty

4. Small packet size

5. Small network size

So, if Modbus is so simple and has been successfully implemented thousands of times over, do we really need another book on Modbus? Well, yes.

We live in a new age. The age of enterprise communications. It's an age where automation and the factory floor are changing in ways that weren't imaginable just a few short years ago. Today, IT and IT technologies are rapidly moving on to the factory floor. The push to totally integrate business from the lowest sensor to the most extensive business system is irreversible. New technologies, new processes, new kinds of organizations and new systems are upon us in ways we've never imagined.

And despite all this, Modbus is still going to be with us. Modbus devices have permeated every kind of automation and will continue to over the next hundred years due to their simplicity and because they're perfect for a lot of simple devices.

This book describes these changes and the role that Modbus will continue to play. It's been successful for forty years now it will continue to be successful over the next forty.

John Rinaldi
June 1, 2015
Paris, France

LEARN MORE An audio interview with the author is available on the resources web page for this book. Just visit either of the following web pages:

http://www.rtaautomation.com/technologies/modbus-rtu/
http://www.rtaautomation.com/technologies/modbus-tcpip/

A LITTLE MODBUS HISTORY

You might call the Modbus protocol the grandfather of industrial networking. It truly is as old as the hills and has the whiskers to prove it. In today's age of Internet connectivity and Web Services, Modbus' unconnected message and simple request-response communication structure is almost quaint. Almost as old as the first Programmable Logic Controller, the Modicon 084, which in those days was called a PC for Programmable Controller.

Modbus is an open standard, meaning that manufacturers can build it into their equipment without having to pay royalties. It is the most pervasive communications protocol in industrial automation, and is now the most commonly available means of connecting industrial electronic devices.

Modbus is used widely by many manufacturers throughout many industries. Modbus is typically used to transmit data from control instrumentation to a logic controller or a system for archiving data. In building automation, for example, temperature and humidity are often communicated to a computer for long term storage. Modbus is often used to connect a supervisory computer with a remote terminal unit (RTU) in supervisory control and data acquisition (SCADA) systems.

Before Programmable Controllers, Control Engineers did relay control. Hardwired relays on the wall acted as the machine logic. There were rooms with walls full of relays, terminal blocks and more wire than you could easily measure.

Figure 1 - Typical Relay Panel

The walls came to be organized with power lines down the sides connecting various control inputs to various kinds of output relays. The structures began to look like ladders so the term "ladder logic" came to represent that kind of control logic.

This wasn't, as you might say, optimum. The problems were many, including:

- A massive amount of time required to change the control logic. To do it properly you moved the control inputs and output relays around. To do it quickly you just rewired it, creating a control system that couldn't be understood.

- These control rooms weren't ventilated properly. Control Engineers at the time didn't have the know-how to make good terminations. Contacts often failed as they became worn or dirty and machine downtime from loose wires was common.

- A complete lack of systems to document the control system.

It was common to spend hours and hours to track down a problem which could be fixed in 10 seconds by wiping off a contact.

This was the era that birthed the PLC. Richard Morley and several associates founded the Modicon Corporation in the 1968. They introduced the first Programmable Controller, the "084," so named because it was the 84th project at Bedford and Associates, the company they had left to found Modicon.

Did You Know?
The first PLC's
were called PC's,
which stood for
Programmable
Controller

Ten years after that first PLC and after a few successors to the 084, Modicon introduced Modbus, the world's first industrial communication network and arguably the most successful. Modbus connected those Modicon PLCs to each other and to remote devices starting a whole new era in factory floor connectivity.

Yet beyond being an originating standard bearer for communication protocols, it also remains relevant to many applications and systems today.

WHY MODBUS HAS FLOURISHED

Modbus, as we learned in the last chapter, is the most pervasive communications protocol in industrial and building automation and the most commonly available means of connecting automated electronic devices.

Why did that happen? Why did Modbus have such an impact on the Industrial Automation industry that it survives to this day as one of the leading industrial networks of the 21st century? There are three primary keys to its success.

Modbus Is an Open Standard

Modicon did not keep the standard proprietary. They released it as a non-proprietary standard and welcomed developers, even competitors, to implement it. They rightly assumed that it would be best for everyone, including them, if Modbus became successful in the marketplace. Because of this thinking, Modbus became the first widely accepted fieldbus standard. In a short time, hundreds of vendors implemented the Modbus messaging system in their devices and Modbus became the de facto standard for industrial communication networks.

Modbus Uses Standard Transports

The transport layer for Modbus RTU commands is also simple to understand. It uses RS485, a differential communication standard which supports up to 32 nodes in a multi-dropped bus configuration. RS485 provided superior noise immunity than the RS232 electrical standard.

Modbus Uses a Simple Protocol

Modbus is very easy to understand. Its primary purpose is to simply move data between a RTU Master device (a Client in Modbus TCP) and an RTU Slave device (a Server in Modbus TCP). There are only two kinds of data to move, registers and coils. Registers are 16-bit unsigned integers. Coils are single bits.

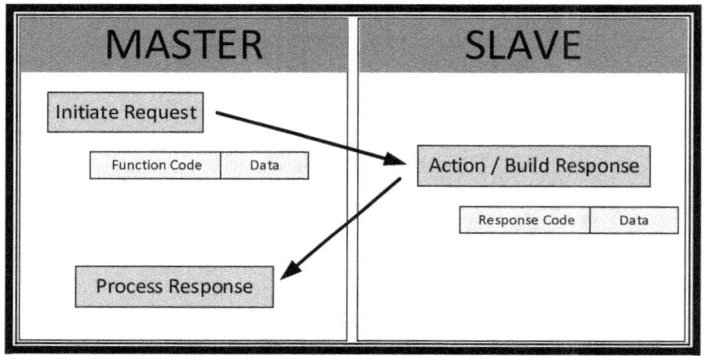

Figure 2 - Master/Slave Modbus Architecture

Modbus uses a very straightforward Request-Response command structure. A Modbus Master requests or sends data to a Slave and the Slave responds. There are simple commands to read a register, read a coil, write a register and write a coil.

Complete simplicity was extremely important when the protocol was released. Remember that microprocessor technology when Modbus was created was not only brand new, but extremely limited. Programmers often had as little as 64 bytes to manage. That's 64 b-y-t-e-s, not 64K or 64Meg. Developing an automation app with just 64 bytes of RAM meant that every single byte was treasured. There just weren't any bytes to frivolously waste.

Modbus fit well in that area. It required little code space, often as little as 1K. RAM varied with the size of your data space. Simple automation devices with little bits of data—imagine a photo eye—could be implemented with hardly any RAM space. These devices could now, for the first time, send their data to a control system as part of a daisy-chained 485 network, avoiding hardwired point-to-point communications.

Did You Know? Before it's downfall, Enron had its own version of Modbus

Another reason Modbus was so successful was the fact that it could be so readily understood by non-programmers. Engineers that built glue machines, meters, measuring devices and such could easily understand the concept of coils/registers and the simple commands to read and write them.

The simplicity of Modbus has been both a blessing and a curse over the years. The simplicity has led to an incredible amount of activity and propagation of Modbus into many different industries around the world. There is probably no product category in the last thirty years that hasn't had an offering without Modbus.

The simplicity of Modbus has also led to many companies expanding the message structure, data representation and transports. Some vendors have imposed any number of advanced structures and data types on the basic Modbus address structure. Others have used Modbus in other ways that go beyond the basic specification.

These implementation extensions are not expressly prohibited by the specification, but don't always make Modbus easily portable to many different applications.

In spite of this, Modbus has gained wide market acceptance wherever Industrial Automation Systems (IAS) or Building Management Systems (BMS) need to communicate with other devices. In fact, Modbus is probably the most implemented automation protocol of all time.

MODBUS DATA REPRESENTATION

In the movie *City Slickers*, the old cowboy Curly (Jack Palance) tells Mitch (Billy Crystal), the secret of life. "One thing. Just one thing," he says—but that you need to figure it out for yourself. And it's what makes life meaningful again for Mitch when he realizes that his wife and children are all that really matter to him.

Well, Modbus isn't the answer to the secret of life. But I can tell you the one pseudo-secret that's the key to understanding Modbus. Actually the key to understanding DeviceNet, PROFIBUS, EtherNet/IP, PROFINET IO and every other protocol. And that's the data representation. If you understand the data representation for a protocol, you're 80% of the way to understanding what that communications protocol does and how it does it.

Like everything else about Modbus, the data representation is pretty simple. In fact, data is represented more simply in Modbus than in any other industrial protocol you'll ever find.

There are only two data types in Modbus: coils and registers.

Coils:

Coils are simply single bits. The bits can be ON (1) or they can be OFF (0). Some coils represents inputs, meaning they contain the status of some physical discrete input. Or they represent outputs, meaning that they hold the state of some physical discrete output signal.

Registers:

Registers are simply 16-bit unsigned register data. Registers can have a value from 0 to 65535 (0 to FFFF hexadecimal). There is no representation for negative values, no representation for values greater than 65535, and no representation for real data like 200.125.

Applications can impose these representations on registers. For example, a register can treat two registers, the first containing 200 and the second containing 125, as 200.125. Or an application can group four registers and place a 64-bit IEEE floating point bit pattern in those registers. Any application can organize and treat register data in any way it may want, but there is no way for any other Modbus device to automatically know what that representation is. A Modbus application that reads registers from a Modbus Slave device must have some prior knowledge of how particular registers are treated to process them correctly.

Registers are grouped into Input Registers and Holding Registers. Like Input Coils, Input Registers report the state of some external input as a value between 0 and 65535. The original intent of an Input Register was to reflect the value of some analog input. It is a digital representation of an analog signal like a voltage or a current. Most Modbus devices today are not I/O devices, and Input Registers simply function identically to Holding Registers.

Holding Registers were originally designed as temporary program storage for devices like Modbus controllers. Today, Holding Registers function as data storage for devices.

Both Modbus registers and coils are addressed with the first register or coil as Address 1 and the last as Address 65536. That means that there can be up to 65536 (10000 Hex) Input Registers, 65536 Output Registers, 65536 Input Coils and 65536 Status Coils, but most devices use far fewer. Often you will find no coils in a device, and sometimes as few as 10 Holding Registers.

A lot of novice Modbus users find the address space notation used in Modbus confusing. Modbus includes the address space type with the index in the address space. The typical address space notation used in Modbus follows:

0x is Status Coil Address Space from 00000 to 065535

1x is Input Coil Address Space from 10000 to 165535

3x is Input Register Address Space from 30000 to 365535

4x is Holding Register Address Space from 40000 to 465535

So, when you see a notation that some value is at Modbus Register 40010, you know that the value is stored in a Holding Register at offset 10, the 11th value, in the Holding Register area of the device. The reason that this is important is that there are specific Modbus functions that operate on specific areas of the address space. There is a read Holding Register command that always reads registers in the 4x address space. There is a Write Single Coil Register that only writes coils in the 0x address space.

What's important to note about this data addressing is that the coil addressing is specifying bit addresses, while register addresses specify 16-bit unsigned integer values. Input Coil address 2 is a bit address. It's the third bit of the coil address space. Input Holding Register address 2 is the address of a 16-bit value. It's the third value of the Holding Register address space.

The commands that operate on these address spaces reflect those data types. A command to read coils specifying a length of 3 is going to return 8 bits, of which only 3 will be valid. Coil Address 3 for a length of 3 returns coils 3, 4 and 5. A command to read registers specifying a length of 3 returns three 16-bit values. It returns the values of registers 3, 4 and 5.

Write commands work in the opposite direction. When writing three coils, you must specify the three bits you want to write as the first three bits of the input data in the command. But when writing three registers, you must supply three 16-bit values.

Other than some awkward notation that is a left over from the era when Modicon Inc. designed the protocol, the Modbus data representation is very simple and straightforward. When a Modbus device is designed, the designer makes a decision of not only how many registers or coils are needed, but also which address space to use (Input Coil, Status Coil, Input Register, Holding Register) and where in that address space to locate those values. A designer may have only 10 bits of coil data and 2 registers. Those bits and registers can be located anywhere in the Modbus address space.

That example points to one of the deficiencies of this data representation. There are no standards. There is no way to communicate to the user of the device any meta-data. If the value of register 40100 is a temperature, there is no standard way to communicate the meaning of register 40100 and how to interpret the data. The user has no information to know if a value of 1001 means 100.1 degrees or 10.01 degrees or 1001 degrees.

And there is no standard regarding device profiles. Another temperature controller may store its temperature at 40200 and use a completely different data interpretation.

The Modbus data representation is simple and almost standards-free. That is both a blessing and a curse to those of us who still use Modbus on a daily basis.

MODBUS TRANSPORT LAYERS

A computer protocol is nothing more than a series of bits with a known pattern that communicate a message from one computer to another computer. There are lots of computer protocols in the world. I doubt that anyone has ever counted them. It would be like going to Laguna Beach in California with the intent of counting the grains of sand on the beach. They are everywhere and there seem to be more every day.

In Industrial Automation there are a number of famous ones. There is Modbus, of course, the subject of this book—along with DeviceNet and EtherNet/IP (two that are used extensively in the US) and PROFINET IO and PROFIBUS (two that are common in Europe.)

These are nothing more than standardized bit patterns that a receiver knows how to decode.

For example, in a different chapter of this book we describe how to read two bytes of a Holding Register. The sequence to do that is to send the following items on the wire:

STATION ADDRESS	FUNCTION CODE	REGISTER ADDRESS	NUMBER OF REGISTERS

Figure 3 - Modbus Message Structure

To read 2 registers from Holding Register 40310 on station 5, the command sequence in decimal would be "05 03 0310 02." Each of those bytes is converted to a bit pattern and each bit is serially transmitted to the other stations on the network. Station 5 realizes the message is for it. It reports a response in the standard Modbus format.

That's all a computer protocol is; a stream of bits that is well-understood by a sending station and a receiving station.

Now, what we've overlooked in this little example is how those bits move from one station to another. There are a lots of ways to do that, including "sneakernet." Sneakernet is where you write down the command at the sending station, walk over to the receiving station and ask for the response, which you then carry back to the original sending station. Not terribly difficult work, but probably pretty boring.

How we move a message from one station to another is called the Transport Layer. Now, the Transport Layer can be pretty complicated, or it could be downright simple like sneakernet. The Transport Layer used by ZigBee communications would be an example of a pretty complicated one. Your message might travel through hundreds of wireless nodes to get from the sender to the destination. Or, if you have a single wire between the sender and destination, it might be as simple as putting a series of voltages on a line connecting two stations. For each bit that is a one, you would raise voltage on the line. For each zero, you would not raise voltage.

You can use any transport you want. The message

contents is what makes a protocol that protocol. The Modbus message sequence is the Modbus message sequence whether you send it on Ethernet, RS232, RS485 or sneakernet. The transport is simply the mechanism you use to move the message from the sender to the receiver.

There are several standard transports used to move Modbus protocol messages: RS232, RS485 and Ethernet. You can use others, but these are the common ones.

RS232

RS232 stands for Recommend Standard number 232. This is the old serial port that we used to find on computers several (okay, more than several) years ago. The full RS-232C standard specifies a 25-pin "D" connector. Now, if we find a serial connector, it uses the 9-pin D type connectors often referenced as a DB9.

RS232 transports bits by driving a voltage potential across two wires, the transmit wire and a ground wire. A receiver senses the potential and records either a one or a zero. There are some synchronizing ones and zeros and some standard bit times that allow both the sending and receiving station to synchronize the transmission and reception.

An important characteristic of RS232 is that it is a single sender, single receiver system. It is electrically not possible for the RS232 electronics to drive a signal to multiple destinations. Since Modbus is mostly designed as a Master sender, sending to multiple Slave receivers, RS232 is only used very rarely as a Modbus transport layer.

Did You Know? The commonly referenced DB9 is technically a DE-9. The E represents the housing size.

RS485

RS485 is a successor to RS232. It works in a similar fashion regarding the synchronizing bits that synchronize the transfer of bits from a sending station to a receiving station. There are, however, two defining characteristics that make RS485 different from RS232. The first is the ability to drive multiple destinations. RS485 transmitters have the ability to electrically signal up to 32 destination devices. That makes RS485 the preferred way to serially transport Modbus messages.

Did You Know? The twisted pair was invented by Alexander Bell in 1881.

The other defining characteristic of RS485 is enhanced noise immunity. RS485 does not use the electrical common as the reference for its electrical signal. Instead, RS485 uses a pair of wires and drives a signal by setting a voltage potential across the pair. By doing that, any environmental electrical noise affects both wires equally and the potential across the two wires isn't changed. This is a vast improvement over RS232, and it has made possible the Modbus communication from a single Master to many Modbus slave devices.

MESSAGE ENCODING

An encoding mechanism describes how bit patterns are formed from the control and data values that are encoded into the packet. Both the sender and the receiver have to use the same encoding to correctly understand the contents of the data. There are two mechanisms for encoding

Modbus messages, ASCII and RTU.

RTU encoding is the newer and much more common encoding mechanism used on Modbus. RTU simply means that values are encoded as standard big-endian binary. That means that in the case of 16-bit values, the Most Significant Byte (MSB) is encoded prior to the Least Significant byte (LSB). An 8-bit value like decimal 41 (29 hex) is encoded simply as 0010 1001. Whereas a 16-bit value like decimal 300 (12C hex) is encoded as 0000 0001 0010 1100. The MSB of 01 is encoded and transmitted prior to the LSB of 2C.

ASCII encoding is a remnant of the old teletype days. It is still used on some very old Modbus equipment. When a data value is ASCII encoded, the ASCII values are transmitted on the wire instead of the binary values. For example, the decimal value 41 is transmitted as an ASCII 4 (34 hexadecimal) followed by an ASCII 1 (31 hexadecimal). The value 41 is transmitted as 3431 or 0011 0100 0011 0001. The 8 bit value 41 is actually sent as 16 bits. This was required in the old teletype days, but not any longer; however, some old Modbus devices can still be found that support it.

ETHERNET (TCP/IP)

Modbus communications over Ethernet is known as Modbus TCP. Modbus took a huge step forward when it began using Ethernet as a transport protocol. Now with Ethernet, Modbus TCP devices can be located miles away instead of hundreds of feet, and with the increased speed of Ethernet, there is much more bandwidth for sending many more messages. You can connect thousands of devices in a Modbus TCP network, not just the 32 like in RS485, and you can also support multiple Masters, not just a single Master.

Modbus TCP is nothing more than the same Modbus protocol we've discussed elsewhere in this book transported over Ethernet. That's actually a misnomer, because Modbus is not traveling in an Ethernet packet. In actuality, the same Modbus message we've discussed is traveling within a TCP packet, which is itself traveling in an IP packet, which is using Ethernet electrical signaling to move a message from point A to point B.

LEARN MORE If you're new to TCP/IP, you can get a TCP/IP Protocol Guide.Just visit either of the following web pages:

http://www.rtaautomation.com/technologies/modbus-rtu/
http://www.rtaautomation.com/technologies/modbus-tcpip/

Let's look at this in a little more detail. Ethernet is a signaling mechanism. It is an electrical standard for moving bits from one station to another station. There is a protocol that is used to do that transfer of a message from one Ethernet station to another Ethernet station. That is the IP, or Internet Protocol. Among other things, the Internet Protocol identifies the sending and receiving stations. If your message, for example, has to be routed through a router, the first step of the transfer is from your sending station to the router. The IP protocol handles that.

Once there, your message has to be processed and the next step of the journey figured out. That is where the TCP protocol comes into play. Within the data of the IP packet is the TCP packet. Among other things, this protocol contains the TCP/IP address of the destination station. From that TCP/IP address, the next stop on the way to your destination is identified and the TCP/IP packet is packed again in another IP packet and sent to that next destination where the process is repeated.

Eventually your message lands in the final destination. The IP packet is unpacked again. The TCP packet is unpacked and your Modbus message, which is the data contents of the TCP packet, is delivered to the Modbus protocol software at that destination.

At that point, the Modbus Master (or Slave) processes that message just as it would any other Modbus message.

There is one difference between the Modbus message on Modbus TCP from the Modbus message on RS485 or RS232. On Modbus RS485 (or RS232), Modbus messages are transmitted with a 16-bit CRC or cyclic redundancy check at the end. This is a bit pattern that is used to validate the message. If any bits have been incorrectly transmitted, the CRC pattern can detect that and signal an error. On Modbus TCP, that functionality and that CRC is discarded as there is a CRC serving the same purpose in the TCP packet.

WIRELESS MODBUS TRANPORT

Wireless in another transport layer that can be used to transport Modbus messages. There are many wireless protocols, but wireless Ethernet or 802.11 is the most common way of moving Ethernet packets through the air.

There is nothing special about moving Modbus messages wirelessly. Since Modbus TCP uses standard Ethernet, wireless is just another kind of "wire." Nothing special is required for wireless Modbus. You can pick any of the standard wireless offerings, including the frequency band (2.5GHz or 5GHz) and how you want the channels to operate.

Sometimes, though, you can use a special wireless device called a Wireless Device Server. This is a device that serves as a serial Modbus RTU Master device to a remote network of serial Modbus Slaves. These devices move Modbus data from these RTU Slaves over the Ethernet network as TCP messages. A complementary device on the receiving end converts the Modbus message back into its original form and sends it out over an RS232 or RS485 electrical interface.

There are several benefits to wireless Modbus messaging:

* Multiple interfaces for ease of integration.

* Simpler and faster maintenance with remote diagnostics and configuration.

* Redundant wireless provides fault-tolerant systems and high-reliability networks.

- Digital status output often provides wireless link indication status for fault notification.

- TCP to RTU conversion enabling Ethernet and serial Modbus devices on the same network.

MODBUS – THE DETAILS

Modbus Protocol Message Structure[1]

The standard structure of a Modbus request and a Modbus response is identical for all forms of Modbus messages. That structure begins with a byte indicating the function the Slave should perform and ends with the last byte of data in the message.

Message requests and responses for both Modbus TCP and Modbus Serial use EXACTLY the same byte sequences for this part of the message. The header and trailer data used by Modbus TCP varies slightly and is described in a later section.

The general components of a Modbus message follow:

Function Code (FC)   The Function Code identifies the request to the Modbus Slave. There are a large number of possible message requests, but about eight that are commonly used. These are the

---

[1] This is an abridged reference to the complete Modbus Specification. See the Modbus Specification for complete details on all Modbus function codes.

THE EVERYMAN'S GUIDE TO MODBUS

	function codes that are detailed in this chapter.
Starting Address	The Starting Address is the index into the data area in the Modbus device. If the function code targets coils, this field specifies the index the coils (bits) of the coil address space. If the function applies to registers, this field specifies the index into the registers for that part of the address space.
	Note: Modbus address spaces are one based—the first register or coil is one. The Modbus protocol is zero based. The first register or coil is zero. The address on the wire is always one less than the address in the Modbus data request.
Bit Length	The number of bits to read or write.
Word Count	The number of registers to read or write.
Byte Count	The number of data bytes of data included in the message request or response.
Response Code	This byte indicates the successful completion of the message request. It is identical to the original message request.
Exception Response (FC)	An exception response is indicated by combining the Response Code of the original Modbus Function Request with 80 hexadecimal. For example, a Modbus exception response to Function Code 3 is 83 hexadecimal. A single data byte value with the Modbus Error Code always follows the exception response byte.

Table 1 - Modbus Message Components

The structure of the most commonly used Modbus Requests and associated Modbus Responses follow:

READ COILS (FC=1)

The Read Coil message is used by a Modbus Master to move one or more coils from a Modbus Slave device to the Modbus Master. Up to 2000 coils (250 bytes) can be read by the Modbus Master in one message.

Read coils (FC 1)	
Request	Byte 0: FC = 01 Byte 1-2: Starting Address (16-bit) Byte 3-4: Bit count (1-2000)
Response	Byte 0: FC = 01 Byte 1: Byte count of response (B=(bit count+7)/8) Byte 2-(B+1): Bit values (least significant bit is first coil!)
Exception Response	Byte 0: FC = 81 (hex) Byte 1: exception code
Example: Read 1 coil at coil address 1 resulting in value 1 Request: 01 00 00 00 01 Response: 01 01 01	

Table 2 – Read Coil Structure

READ DISCRETE INPUTS (FC=2)

The Read Discrete Inputs message is used by a Modbus Master to move one or more Input Coils from a Modbus Slave device to the Modbus Master. Up to 2000 coils (250 bytes) can be read by the Modbus Master in one message.

Read Discrete Inputs (FC 2)	
Request	Byte 0: FC = 02 Byte 1-2: Starting Address (16-bit) Byte 3-4: Bit count (1-2000)
Response	Byte 0: FC = 02 Byte 1: Byte count of response (B = (bit count+7)/8) Byte 2-(B+1): Bit values (least significant bit is first coil!)
Exception Response	Byte 0: FC = 82 (hex) Byte 1: exception code
Example: Read 1 Input Coil at coil address 1 resulting in value 1 Request: 02 00 00 00 01 Response: 02 01 01	

Table 3 – Read Input Coil Structure

READ HOLDING REGISTERS (FC=3)

The Read Holding Registers message is used by a Modbus Master to move one or more Holding Registers from a Modbus Slave device to the Modbus Master. Up to 125 registers can be read by the Modbus Master in one message.

Read Holding Registers (FC 3)	
Request	Byte 0: FC = 03 Byte 1-2: Starting Address (16-bit) Byte 3-4: Word count (1-125)
Response	Byte 0: FC = 03 Byte 1: Byte count of response (B=2 x word count) Byte 2-(B+1): Register values
Exception Response	Byte 0: FC = 83 (hex) Byte 1: exception code
Example: Read 1 Holding Register at register address 1 resulting in value 1234 Hex Request: 03 00 00 00 01 Response: 03 02 12 34	

Table 4 – Read Holding Registers Structure

READ INPUT REGISTERS (FC=4)

The Read Input Registers message is used by a Modbus Master to move one or more Input Registers from a Modbus Slave device to the Modbus Master. Up to 125 registers can be read by the Modbus Master in one message.

Read Input Registers (FC 4)	
Request	Byte 0: FC = 04 Byte 1-2: Starting Address (16-bit) Byte 3-4: Word count (1-125)
Response	Byte 0: FC = 04 Byte 1: Byte count of response (B = 2 x word count) Byte 2-(B+1): Register values
Exception Response	Byte 0: FC = 84 (hex) Byte 1: exception code
Example: Read 1 Input Register at Register Address 1 resulting in value 1234 Hex Request: 04 00 00 00 01 Response: 04 02 12 34	

Table 5 – Read Input Registers Structure

WRITE SINGLE COIL (FC=5)

The Write Single Coil message is used by a Modbus Master to set one coil in the Output Coil Address Table in the Modbus Slave device.

Write Single Coil (FC 5)	
Request	Byte 0: FC = 05 Byte 1-2: Starting Address (16-bit) Byte 3: = FF to turn coil ON, = 00 to turn coil OFF Byte 4: = 00
Response	Byte 0: FC = 05 Byte 1-2: Starting Address (16-bit) Byte 3: = FF to turn coil ON, = 00 to turn coil OFF (echoed) Byte 4: = 00
Exception Response	Byte 0: FC = 85 (hex) Byte 1: exception code
Example: Write 1 Coil at Coil Address 1 to the ON value. Request: 05 00 00 FF 00 Response: 05 00 00 FF 00	

Table 6 – Write Single Coil Structure

WRITE SINGLE REGISTER (FC=6)

The Write Single Registers message is used by a Modbus Master to move one Register value from the Modbus Master to a Modbus Slave device.

Write Single Register (FC 6)	
Request	Byte 0: FC = 06 Byte 1-2: Starting Address (16-bit) Byte 3-4: Register value
Response	Byte 0: FC = 06 Byte 1-2: Starting Address (16-bit) Byte 3-4: Register value
Exception Response	Byte 0: FC = 86 (hex) Byte 1: exception code
Example: Write 1 Holding Register at register address 1 with a value 1234 Hex Request: 06 00 00 12 34 Response: 06 00 00 12 34	

Table 7 – Write Single Register Structure

WRITE MULTIPLE COILS (FC=15)

The Write Multiple Coil message is used by a Modbus Master to set one or more coils in a Modbus Slave device. Up to 2000 coils can be written by the Modbus Master in one message.

Write Multiple Coils (FC=15)	
Request	Byte 0: FC = 0F (hex) Byte 1-2: Starting Address (16-bit) Byte 3-4: Bit count (1-2000) Byte 5: Byte count (B = (bit count + 7)/8) Byte 6-(B+5): Data to be written (least significant bit = first coil)
Response	Byte 0: FC = 0F (hex) Byte 1-2: Starting Address (16-bit) Byte 3-4: Bit count
Exception Response	Byte 0: FC = 8F (hex) Byte 1: exception code
Example: Write 3 coils at coil address 1 to value 0, 0 and 1 Request: 0F 00 00 00 03 01 04 Response: 0F 00 00 00 03	

Table 8 – Write Multiple Coil Structures

WRITE MULTIPLE REGISTERS (FC=16)

The Write Multiple Registers message is used by a Modbus Master to move one or more Registers from the Modbus Master to a Modbus Slave device. Up to 125 registers can be written by the Modbus Master in one message.

The Write Multiple Registers command is often used in place of the Write Single Register. Some Modbus Slaves can only support the Write Multiple Registers command.

Write Multiple Registers (FC 16)	
Request	Byte 0: FC = 10 (hex)
	Byte 1-2: Starting Address (16-bit)
	Byte 3-4: Word count (1-125)
	Byte 5: Byte count (B = 2 x word count)
	Byte 6-(B+5): Register values
Response	Byte 0: FC = 10 (hex)
	Byte 1-2: Starting Address (16-bit)
	Byte 3-4: Word count
Exception Response	Byte 0: FC = 90 (hex)
	Byte 1: exception code
Example: Write 1 Holding Register at reference 1 with a value 1234 Hex	
Request: 10 00 00 00 01 02 12 34	
Response: 10 00 00 00 01	

Table 9 – Write Multiple Registers Structure

READ/WRITE MULTIPLE REGISTERS (FC=23)

The Read/Write Multiple Registers message is used by a Modbus Master to combine a write operation with a read operation. Registers can both be read from the Modbus Slave device and written to the Modbus Slave device.

Read/Write Multiple Registers (FC 23)	
Request	Byte 0: FC = 17 (hex) Byte 1-2: Starting Address (16-bit) for read Byte 3-4: Word count for read (1-125) Byte 5-6: Starting Address (16-bit) for write Byte 7-8: Word count for write (1-125) Byte 9: Byte count (B = 2 x word count for write) Byte 10-(B+9): Register values for write
Response	Byte 0: FC = 17 (hex) Byte 1: Byte count(B = 2 x word count for read) Byte 2-(B+1) Register values
Exception Response	Byte 0: FC = 97 (hex) Byte 1: exception code
Example: Read two registers at address 1 and write the value 123 hex to register address four with a read value of 0004 5678 hex. Request: 17 00 00 00 02 00 03 00 01 02 01 23 Response: 17 04 00 04 56 78	

Table 10 – Read/Write Register Structure

You can get a link to YouTube
Videos introducing Modbus RTU and Modbus TCP on
the resources web page for this book. Just visit either of
the following web pages:

http://www.rtaautomation.com/technologies/modbus-rtu/
http://www.rtaautomation.com/technologies/modbus-tcpip/

Modbus Message Encoding

Serial Modbus supports two encoding standards: RTU
and ASCII. ASCII is the older, seldom used standard where
every byte is encoded as two ASCII characters. The value
14 hexadecimal, for example, is encoded as the ASCII "1"
(31 hexadecimal) followed by the ASCII "4" (34
hexadecimal). With the extra memory, processing time and
lower network bandwidth of Modbus ASCII, it is only
found in very old equipment.

The RTU encoding standard is the one used for both
Serial Modbus and Modbus TCP. In RTU, a byte is
encoded as its binary equivalent.

Modbus RTU Station Identification

Serial Modbus Slave devices are identified by a station
number which precedes the general message structure.
Generally, up to 32 stations are supported as that is the
limit imposed by most RS485 serial drivers. There is no
software limit to the number of stations that could be
supported. Valid Slave addresses are assigned in the range
of 1 to 247 with station number 0 reserved for broadcast
messages, messages processed by all Slaves.

Modbus TCP Station Identification

Devices are not identified in the same way on Modbus TCP. On Modbus TCP, Server devices are identified by their TCP/IP address. A Modbus TCP Client device issues a connection request and creates a TCP connection with each Modbus TCP Server device. All messages are then transferred over that TCP connection.

In some applications, a TCP/IP address can represent a number of devices on some sub-network. In those cases, a secondary ID field is used to identify the node on the sub-network.

Exception Reporting

Modbus devices report exceptions using the exception response (80Hex plus the Function Code) and an error value. Some of the more common error values are:

01 Illegal Function: The function code received in the query is not allowed. It is either not supported, or the Slave is in a state in which that function code cannot be processed. For example, asking a Slave to return the values of a register prior to it being configured could result in this error code.

02 Illegal Data Address: The data address received by the Slave is not valid for this Slave. This typically means that the Master (or Client) is asking for one or more registers or coils outside the range of registers and coils defined for the address table of the slave. More precisely, the combination of Starting Address (16-bit) and transfer length results in one or more coils that are beyond the last register or coil of the address table.

03 Illegal Data Value: A value contained in the query data field is not allowed. This always indicates a fault in the structure of the Modbus message or invalid parameter like a length of 50,000. It never means that the data value is invalid, as Modbus is unable to validate the value of a register.

04 Slave Device Failure: An unrecoverable error occurred while the Slave was attempting to perform the requested action.

10 Gateway Path Unavailable: A specialized use in conjunction with gateways. This indicates that the gateway was unable to allocate an internal communication path from the input port to the output port for processing the request. Usually means the gateway is misconfigured or overloaded.

11 Gateway Target Device Failed to Respond: Specialized use in conjunction with gateways, this indicates that no response was obtained from the target device. Usually means that the device is not present on the network.

Quick Tip: In practice, exceptions 1, 2 and 3 are the only exceptions you are likely to run into.

MODBUS TCP STRUCTURE

Even though the Modbus message structure doesn't change in Modbus TCP, there are a few differences between Modbus TCP and serial Modbus.

In Modbus, there is a Device ID or Station ID that identifies the Slave device. Modbus Slaves ignore messages that have a Device ID that is different than their configured ID.

Serial Modbus messages also contain the two CRC bytes at the end of the message.

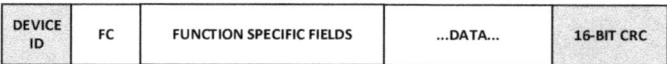

Figure 4 - Serial Modbus Message Structure

Modbus TCP messages have extra fields that precede the Modbus message.

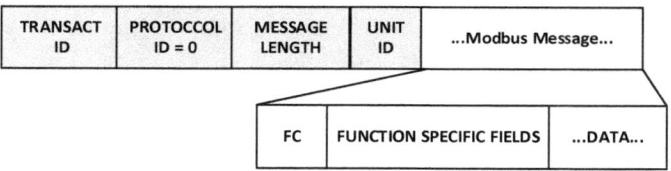

Figure 5 - Modbus TCP Message Structure

Note from Figure 4 and Figure 5 that the core message structure for both protocols is identical. The serial Modbus structure contains identification and error checking fields while the Modbus TCP message structure contains additional fields which are largely unused.

The transaction ID field of the Modbus TCP message is simply returned in the Modbus TCP response. The protocol ID is zero and doesn't change. The Message Length is the length of the entire Modbus TCP message. And the Unit ID is used to identify a target on a secondary network in gateway applications.

Other than these special fields, the core Modbus message following the Modbus TCP header remains identical to serial Modbus.

MODBUS FOR SOFTWARE
DEVELOPERS

A long time ago I was a software developer and I still have a soft spot in my heart for all of you that plug away, day in and day out, trying to write great software. If you are a developer and need to implement the Modbus protocol on a device, then this chapter's for you.

There are a number of questions you need to answer before ever picking up a keyboard or even surfing a website.

#1 – Are You Using Modbus TCP or Modbus Serial?

You may be tempted to immediately say you want to do Ethernet and Modbus TCP, but let's look at this a little closer. There is an advantage to Modbus Serial. If you have a low-resourced device that needs to communicate bytes of information sporadically, Modbus Serial may be the right choice. If all you have is a UART and only a few K of code space, it may be the right choice. If your Modbus Master device is located nearby (within 1000 feet) you're all set. No reason to look any further.

Now, there may be good reasons to use Ethernet and Modbus TCP. You may have lots and lots of devices on the network. The devices may be miles away. You may have lots of registers and coils to transfer. You may want to easily transition to a wireless 802.11 solution in the future. You may want to allow more than one client to access your data. There's lots of good reasons for Modbus TCP.

The downside is that you absolutely must have a TCP/IP stack. A TCP/IP stack has a cost associated with it. You'll have to buy one if you want a reliable one, and that stack is going to take up resources. You'll also need the MacID for Ethernet and an Ethernet PHY. Besides the software cost, you have more costly hardware.

My recommendation: Use Modbus Serial if you can.

#2 – Are you the Master device or the Slave device?

Modbus Slaves (Serial RTU) and Modbus Servers (Ethernet) are *end devices*. They respond to the requests of a Modbus Master (Serial RTU) or a Modbus TCP Client (Ethernet). Most automation devices that are not controllers are going to be end devices.

#3 –RTU or ASCII encoding?

This one is easy. Do RTU. Don't even consider implementing ASCII. It's not the 15th century.

#4 – What is Your Data Representation?

This is a very important question. What data do you want to expose to a Modbus Master or Client? List everything you want to expose and then decide for each item if it is going to be in a register or a coil. If you have bit type data, say a series of 12 light switches, you should represent that data as coils.

Once you've assembled your list, you'll want to expand it. What do I mean by that, you might be thinking? I mean take everything you have and figure out other ways for the customer to access that data.

For example, go back to our example of the 12 light switches. In addition to making them available as coils, you should also make them available in a register. The customer can choose to read the 12 coils in a read coil command or read the register where 12 of the 16 bits in the register are identical to those coil bits. That lets the customer have the data in the format in which they can most easily deal with it. If they have an application where they will be reading switch 5 all the time, they can just get that as a coil. If they have an application where they want to record the states every minute, they'll probably just read the register and store it.

If you have analog data, you can also store it in multiple registers. If the value is 1902.54, you might have the following register setup:

Register 40010 – Integer value (1902)
Register 40210 – Decimal value x 10 (19025)
Register 40310 and 40311 – Decimal value x 100 (190250)
Register 40410-13 – IEEE Floating point value for 1902.54

With this strategy, your customer can now access the value in the format that is most convenient and not have to do much manipulation on the data type.

We often do just that sort of thing with temperatures, flows and other kinds of data that have various kinds of units associated with it. We'll have some registers with the Centigrade values and some others with Fahrenheit values. We assign one register as a speed in gallons per minute, and another register as the speed in liters per minute.

The general rule is to always give the customer more data formats than they need. Since only the data requested is passed, there are no bandwidth or network load penalties. You can present as much data in as many ways you as you see fit, and it's up to the customer to determine what they want to access.

#5 – Freeware, Shareware, Purchased Source Code or Writing It Yourself?

This is a big one. I know. I understand. You really want to do this yourself. You're a software developer. That's what you do for a living. Why buy code when you can write it yourself? That's fine if you have the time. As you saw in the Modbus Protocol chapter, it's not rocket science. It's certainly something that you can do.

But what I would suggest is that if you want to use unsupported software, just get one of the freeware packages. There isn't much to be gained from writing it yourself. It's not a big challenge. You won't get any awards or pats on the back from other software developers for writing Modbus. In fact, they'd probably stare at you like you were nuts.

If you do get freeware, check out the source. Make sure that it's good quality. Does it support coils and registers? Does it support all the function codes you'll need? That's probably the main problem with the various types of freeware.

All in all, I'd suggest you buy source code. It's not expensive at all and you always have the vendor to rely on if there is some kind of problem. The vendor will know the kinds of problems that new implementations have, the vendor will know how to troubleshoot an application, and the vendor will have access to tools and resources that you might need.

6. How are you going to test?

Once you have source code in hand, you'll want to build a test network. If you're building a Slave or Server device, you'll want to have a Master and one or more other Servers. If you are building a Master, you'll want to have a bunch of

Servers. It's always best to test a device on a network with at least four or five other devices.

If you're doing Modbus serial, you'll want to have a mechanism for capturing the serial stream. There are devices that can do this or you can simply create a 485 device that listens and reports everything it sees on the wire.

If you're doing Modbus Ethernet (Modbus TCP) you'll want to have the Wireshark data capture tool (https://www.wireshark.org/). Wireshark gives you the ability to see all the packets on the network. It will even decode them and tell you that this command is a Read Register of this address and this length.

You should also get your code certified. See the chapter on certification for that.

And finally, once your device is complete, list it on the http://Modbus.org website where other people can find it and possibly contact you about using it.

GOOD LUCK with your development!

MODBUS ROUTERS

If you have used Modbus for any time at all, you've run into one or more of the following situations:

- You have a Serial Modbus Master that needs to read or write register or coil data in a Modbus TCP Server

- You have the opposite situation: a Modbus Client that needs to read or write register or coil data in a serial Modbus RTU Slave

- You have both a Serial Modbus Master and a Modbus TCP Client and you need to move Modbus register and coil data from one to the other

All these situations are common and require the use of a Modbus Router. A Modbus Router, if it's implemented properly by the vendor, can remove the transport layer restrictions between Modbus Ethernet Clients and serial Slaves and Modbus Serial Masters and Ethernet Servers. With a router, a Modbus RTU Slave can be equally responsive to a Modbus Client as it is to a serial Modbus

RTU Master. A Modbus TCP Server can be equally responsive to a serial Modbus RTU Master as it is to a Modbus TCP Client.

Let's look at an actual example from an article in the Modbus IDA Newsletter.

ELECSAN S.A. is one of those little boutique companies you typically find in places like Mansfield Heights Michigan, Akron Ohio or Sun Prairie Wisconsin. Like many other tiny brother and sister controls companies, it's about what you would expect; a location in a former fish factory smelling of rockfish, halibut and sea bass and a couple of guys who left the big company behind to strike out on their own.

What you don't expect is to find that those guys aren't American. They're Spanish. And their company isn't in some suburb of Detroit or Cleveland or even Houston. It's in the sleepy little town of Sabadell, 20km north of Barcelona in Spain's beautiful and historic Catalonia district. For more than

Figure 6 - Elecsan Headquarters

fifteen years, Joan Ramon and his team of specialists have supplied the local and international market with counters, timers, indicators, text displays and other types of off-the-shelf and custom controls, much of it communicating over Modbus networks.

With a lot of their Spanish customers requesting more and more remote monitoring and control from Elecsan's control products, Joan Ramon had to face a significant problem. An awful lot of the controls Elecsan had delivered over the years were Modbus

RTU, not Modbus TCP, and a large part of it was point-to-point RS232. After all, when you're a couple of young guys starting out and what you know is control systems, not networking, you tend to deploy some stuff that comes back to bite you. Even those of us with "PE" after our names and years of experience make mistakes that make us look foolish years later.

But they long ago picked Modbus RTU and it's worked well for them. Without a massive effort to replace all those systems in a lot of remote places on this planet, it was going to be tough to find a way to provide the kind of remote monitoring that their customers needed.

Challenge

So Elecsan experimented. They tried serial tunnels and virtual com ports – linking remote Modbus RTU masters with various RTU slaves in other parts of the world. If they set the right message timeouts and prayed to Saint Jordi (George), the patron saint of Catalonia, sometimes things worked out OK. Mostly they didn't. Framing timeouts, message timeouts, lost messages...etc. Sometimes it worked all right, but even if they had mistranslated their customers' Spanish requirements, it didn't read "the system shall work some of the time."

Lots of workarounds were discussed. Most would have been disruptive and expensive, including code changes to all Elecsan industrial controls at customer sites and changes to Elecsan software, addition of computers at remote sites, additional software, and even renting a remote cloud service provider. Money, time and disruption to their customers' business all played a role.

Solution

Joan Ramon, like any good engineer, worked his contacts. Having successfully worked with our team at Real Time Automation several times over the years, he figured that we might have a solution. And we did. Our new BFR3000 Modbus Router could easily facilitate the remote monitoring without changing any of the pre-existing industrial controls or computer programs and while meeting all Modbus TCP and RTU protocol requirements for reliable communications.

Figure 7 - Elecsan Remote Monitoring Architecture

This application featured the use of Modbus RTU , Modbus TCP/IP, serial RS232, Ethernet and wireless technology. At the remote site, Joan Ramon used the BFR3000 Modbus Router in its TCP Server-to-RTU Master Mode. The BFR300 became the Modbus Master at the remote site and provided a TCP Server back to Barcelona that Joan Ramon could access over the Internet. For systems in Spain that had a Modbus Master, Joan Ramon simply used a BFR3000 in RTU Slave-to-TCP Client mode enabling that RTU Master to access the remote slaves as if they were on the same RTU network.

With three operating modes, a friendly user interface, multiple serial interfaces and the ability to connect TCP Clients to RTU Masters, the BFR3000 became Joan Ramon's go-to device for remote Modbus communication.

Legacy Opportunities

The opportunity to increase effectiveness and efficiency of industrial operations with the application of remote monitoring, updating, and troubleshooting of control and automation systems is huge. There is a large installed base of legacy control and automation systems that are still effective and efficient, but the scarcity of experienced people to troubleshoot and keep controls up to date is becoming a challenge. This has always been a challenge, since having an expert on staff for all controls and automation in an industrial location is a large overhead cost. Even if people are available that had training on a particular control, they seldom use this information and have to refresh their knowledge to "get up to speed." Meanwhile, the operations suffer loss of production and/or downtime. The ability to have a remote expert that can update, troubleshoot, and maintain controls and automation solves this problem.

Replacing existing installed control or automation systems simply to add remote capabilities is expensive and disruptive. Remote features can be added to the majority of existing control and automation systems without disturbing them by using a range of methods. Users have large investments in their systems, and adding remote monitoring increases the effectiveness and value of these systems immediately with minimal investment. The opportunity is for manufacturers to achieve more speed, agility and efficiency, leading to a competitive advantage.

The simplicity of a Modbus Router is really attractive. An application can have some serial Modbus RTU Slaves located locally to a serial RTU Master and some Modbus TCP servers located remotely. The Modbus TCP Servers could be remotely located on a wired network or even farther away on a wireless link. And all pretty seamlessly. It's a solution that many Modbus users, like Elecsan, have found very attractive.

SHAMELESS PLUG

"We like the BFR3000 router from Real Time Automation. It was very easy is to use for the final client, and we will recommend it to everyone that needs an Internet connection to their machines." Joan Ramon, Elecsan

For more information on the RTA Modbus Router please contact Real Time Automation or visit: http://www.rtaautomation.com/product/modbus-router/.

THE MODBUS ORGANIZATION

Trade organizations are always helpful, both to the novice and the experienced professional. Personally, I belong to countless trade organizations. The annual events they put on are usually informative. Networking with other professionals using the same technology has its benefits. And if you're a novice, it is the best place to find seasoned professionals who can give you the one or two pieces of advice that will save you countless hours of frustration.

The Modbus Organization (http://www.modbus.org/), based in Hopkinton, Massachusetts, is the trade organization for everything Modbus. It is a non-profit group of users and vendors that work together to set standards, advance Modbus technology and promote its use in applications throughout the world.

The Modbus Organization focuses in the following areas:

Technology: The Modbus Organization is the arbiter of what Modbus is, how it works and the standards for products. The organization works to not only promote the technology, but to make sure it is implemented correctly and in a fashion that promotes interoperability among supplier products. The Modbus Organization is a clearing house for products, technical information, tools and services that are useful to the Modbus community. Visitors to the organization's website can find a list of resources to assist them in the development and use of Modbus.

Promotion: The Modbus Organization promotes the adoption of Modbus Technology by publishing a newsletter highlighting vendor products and projects, exhibiting at trade shows where technology users gather, and holding seminars and other events to train people on the use of Modbus.

Standards: The Modbus Organization sponsors certification efforts to ensure that Modbus products correctly meet the standards specified in the Modbus specification.

The stated goals on the Modbus Organization website are:

- Participate in standards activities worldwide.
- Lead the evolution of the Modbus protocol and its variants.
- Encourage and assist the use of Modbus across a broad spectrum of physical layers and transmission media.
- Maintain and evolve a conformance testing program to insure greater interoperability of Modbus devices.
- Provide information to users and suppliers alike to help them be successful in their use of Modbus.
- Engage in educational and promotional efforts including trade shows, newsletters, this website, and other outreach activities.

Participation in the Modbus Organization is open to product vendors, system integrators, distributors, machine builders and end users who are developing Modbus products or using Modbus technology.

There are significant advantages to belonging to the Modbus Organization:

1. Product listings on the Modbus Organization website. Vendors can list the features and benefits of products. End users, distributors and integrators can quickly and easily find them.
2. The Modbus newsletter provides interesting information on the new products, seminars and latest developments in the industry.
3. Technical training programs at a reduced rate.
4. The opportunity to participate in technical committees that influence the future of the technology.

Members of the Modbus Organization can use the Modbus member logo on their company literature to signify their participation in the leading organization for Modbus technology.

Figure 8 - Modbus Organization Membership Logo

MODBUS CERTIFICATION

If you're a device manufacturer and you've ever developed an EtherNet/IP Adapter, BACnet/IP Server or a PROFINET IO Slave device, you are cognizant of conformance testing. These organizations are sticklers for making sure that products are not released to the market without validating that they have the minimal required functionality. This is typical of most industrial protocols. Most trade organizations specify a protocol test that has to be successfully executed to get a certification statement or logo that a device vendor can use to indicate that the device conforms to the specification of the technology.

This is, of course, extremely important to a lot of very big, very important customers like automotive manufacturers. They don't want to spend their time sorting out the whys and wherefores about why product X doesn't work well with product Y. In fact, it's absolutely the last thing in the world that they want to be doing. They want to buy a certified product specifically so they won't be fighting that battle. They know that there is high probability that your certified product is going to perform correctly in their application.

Modbus has been a little late to this party. For many

years, there was no conformance test for Modbus products. A developer would create something, test it in some fashion, and ship it. It was left to the user to figure out what worked and what didn't work.

In the time since, the Modbus Organization has remedied this problem. They not only established a test procedure, they contracted with the University of Michigan to implement it. You now have two options for certification of your Modbus Slave device or Modbus TCP Server device.

One, you can self-certify. That means that you download the test and execute it yourself. You pledge that your products meets the minimum functionality of a Modbus device as specified by the test.

Two, you can purchase testing from the Modbus Organization. The test lab at the University of Michigan will perform the test for you and validate your device. That, of course, is going to cost you some cash and some time. But you get a second set of eyes and third party testing.

You can find out all about testing, including what tests to perform and what options are available, by visiting the Modbus Organization certification web page at http://www.modbus.org/certification.php.

If you visit that site, you will notice that only your Slave or Server device can be tested. There is no test for Master or Client devices. This is similar to some other protocols. The Controller side of many programs cannot be certified through the trade organization's test facility. The reason is that it is much more difficult to validate the operation of the Master side and would be so involved with the specifics of the Master operation that it would be impractical.

You will also note on that site that only Modbus TCP Server devices can be tested. The Modbus Conformance test is designed for Modbus TCP Ethernet devices, not Modbus Serial devices. To test a Modbus Serial device, you must use a gateway that can connect your Modbus Serial device to a Modbus TCP Client. Any Modbus Router can be used for the test, but the Modbus Organization recommends a product from Schneider Electric. I recommend the BFR3000, the router from Real Time Automation (*Yes, that is another shameless plug*).

MODBUS PLUS

You would think that if a technology had a name like Modbus Plus, it would be related to Modbus. Maybe another kind of platform or another transport layer. You'd think that maybe it would be something useful to you as you move up from serial Modbus technology.

Well, if you thought any of those thoughts, you'd be wrong. Modbus Plus is not really related to Modbus at all. It uses the same data representation, but it's an entirely different protocol built around a token passing scheme by Schneider Electric some years ago.. At that time, DH485, another token passing scheme from Allen-Bradley was popular. I don't have definitive proof, but I believe Modbus Plus was created in part to counter that technology from Allen-Bradley.

Modbus Plus never achieved much traction in the marketplace. It did achieve some good penetration in the Schneider Electric customer base. The specification for it was always kept proprietary. The units were typically built around a proprietary chip set.

Modbus Plus was a pretty high-speed network in its day. At one megabaud, it far surpassed the networks that were available to use for serial Modbus. It supported both peer-to-peer and master/slave type communications. It was mostly used by Modicon PLCs, operator interfaces, PCs and other Schneider Electric data sources to communicate over a twisted-pair cable.

JOHN RINALDI

MODBUS & OTHER PROTOCOLS

In order to be useful, Modbus register data often needs to move someplace else, so it can be combined with other data, archived or acted upon in some fashion. Often you'll want to move Modbus data to a Siemens PLC or a Rockwell Controller. That means you'll need to get your Modbus data onto PROFINET IO, EtherNet/IP, PROFIBUS DP or DeviceNet. Before we talk about how to do that, let's do a little refresher on those technologies.

PROFINET IO

PROFINET IO is an Ethernet application layer protocol for industrial automation applications. Built on standard Ethernet technologies, PROFINET IO uses traditional Ethernet hardware and software to define a network that structures the task of exchanging data, alarms and diagnostics with Programmable Controllers and other automation controllers. PROFINET IO is the Ethernet application layer standard for Siemens PLCs.

PROFINET IO is very similar to PROFIBUS on Ethernet. While PROFIBUS uses cyclic communications to exchange data with Programmable Controllers at a maximum speed of 12 megabaud, PROFINET IO uses cyclic data transfer to exchange data with Programmable Controllers over Ethernet. As with PROFIBUS, a Programmable Controller and a device must both have a prior understanding of the data structure and meaning. In both systems, data is organized as slots containing modules with the total number of I/O points for a system composed of the sum of the I/O points for the individual modules.

PROFINET IO devices are represented on the network as an I/O rack with slots, subslots and channels. Subslots are subcomponents of a slot. Each subslot is assigned a number of I/O points or channels. A channel is the PROFINET IO term which refers to one physical discrete input, discrete output, analog input or analog output. A device can have almost any number of slots, subslots and channels.

Modules are specific functional components that can be associated with a slot. Modules can be virtual or real. A module, real or virtual, must be plugged into a slot before the I/O device goes online. The module gives the slot a specific identity. For example, a 16 discrete input module gives the slot a 16 discrete input identity. In the same way that modules provide identity to slots, sub-modules provide subslot identity. The 16 discrete input module can be composed of one 16 discrete input sub-module, two 8 input sub-modules or four, 4 input sub-modules.

In an IO-Controller like a Siemens 317 Programmable Controller, the I/O points or channels for the subslots assigned to that Programmable Controller is collected and that data group forms the I/O data image that is transferred between the IO Device and the IO-Controller. For example, if you have a PROFINET IO device with four input slots (16 inputs each) and two output slots (16 outputs each), the I/O image transferred between the controller and device is 4 bytes in the direction of the Programmable Controller and 2 bytes in the direction of the

IO Device.

PROFIBUS DP

PROFIBUS is the older cousin of PROFINET IO. It was born out of a combined push by the German government, German companies, and other industry leaders in the late 1980s for a networked automation solution. Their effort created an automation solution that is still very viable today. In 1993, the group introduced the PROFIBUS DP (Decentralized Periphery) standard. This new version featured more simplicity, including easier configuration and faster messaging.

PROFIBUS International now manages the technology. It's now known as the PI and is the largest Fieldbus user association in the world. It educates users on PROFIBUS, assists with quality assurance, sets standards and develops new PROFIBUS technologies.

Unlike PROFINET IO, PROFIBUS is a serial fieldbus technology. It's a very high speed serial implementation, a sort of souped-up RS485, but still a serial communication technology. Devices on the system connect to a central line. Devices on PROFIBUS DP communicate information in a very high speed (12 megabaud) and very efficient manner, but also go beyond just transferring IO. PROFIBUS devices can also participate in self-diagnosis and connection diagnosis.

PROFIBUS uses the same data representation as was discussed for PROFINET IO. Data is represented on the network as an I/O rack with slots, subslots and channels. Subslots are subcomponents of a slot. Each subslot is assigned a number of I/O points or channels. A channel is the same term used on PROFIBUS to refer to one physical discrete input, discrete output, analog input or analog output. A device can have almost any number of slots, subslots and channels.

EtherNet/IP

EtherNet/IP is the Ethernet application layer protocol that is used by Rockwell Automation. EtherNet/IP uses the tools and technologies of traditional Ethernet. EtherNet/IP uses all the transport and control protocols used in traditional Ethernet including the Transport Control Protocol (TCP), the Internet Protocol (IP) and the media access and signaling technologies found in off-the-shelf Ethernet interface cards. Building on these standard PC technologies means that EIP works transparently with all the standard off-the-shelf Ethernet devices found in today's marketplace. It also means that EIP can be easily supported on standard PCs and all their derivatives. Even more importantly, basing EIP on a standard technology platform ensures that EIP will move forward as the base technologies evolve in the future.

EtherNet/IP has two device types: the Scanner and the Adapter. The Scanner, almost always a PLC, makes connections to Adapters and writes Outputs to them. The Adapter accepts connections and delivers Inputs to the Scanner.

EtherNet/IP uses two message types—Explicit and Implicit. Explicit messages are message/response oriented and used to asynchronously access data in an Adapter device. Implicit messages (I/O messages) are used for control. Inputs cyclically flow from an Adapter to a Scanner. Outputs cyclically flow from the Scanner to the Adapter.

I/O messages are simply a series of bytes understood by both the Controller and the Adapter. There is a series of data bytes that cyclically move from the Scanner to the Adapter and another series of data bytes that cyclically move from the Adapter to the Scanner. There is no header or identifying information on this data. Both the Controller and Adapter have to have a prior understanding of the contents of those buffers.

DeviceNet

DeviceNet is a low-level industrial application layer protocol for industrial automation applications. DeviceNet connects simple industrial devices (sensors and actuators) with higher-level devices such as Programmable Controllers. Built on the standard CAN (Controller Area Network) physical communications standard, DeviceNet uses CAN hardware to define an application layer protocol that structures the task of configuring, accessing and controlling industrial automation devices.

Like EtherNet/IP, DeviceNet is based on the Communications and Information Protocol (CIP), a communications protocol for transferring automation data between two devices. Where EtherNet/IP is a combination of the CIP Protocol and Ethernet, DeviceNet is a combination of the CIP Protocol and the CAN Physical Layer. Just like in EtherNet/IP, every network device represents itself as a series of objects. Each object is simply a grouping of the related data values in a device. For example, every CIP device is required to make an Identity Object available to the network. And just like EtherNet/IP, DeviceNet uses the same mechanism for transferring I/O data back and forth between a controller and a DeviceNet device.

Modbus and PROFINET IO/PROFIBUS DP

Communicating between Modbus and PROFINET IO requires two things. One, mapping of the Modbus register and coil data into the module, slot and channel structure of PROFIBUS and PROFINET IO. And two, connecting the physical layers of PROFINET IO and PROFIBUS to the serial physical layer of Modbus or Modbus TCP.

Data mapping is not difficult, but it is sometimes

laborious. Discrete Input or Output data points in a discrete I/O module can be represented as a coil data in Modbus. Analog input or output data points in an analog input or output module can be represented as register data in Modbus. And once mapped, the data representations must be documented in a GSD (General Station Description) file for PROFIBUS and a XML version of that file, GSDXML, for PROFINET IO.

You can find gateways that can move data from Modbus Slaves or Modbus Servers and represent that data as PROFINET IO Server data. RTA offers several gateways[2] that you can use to move data between a Siemens Controller and a Rockwell PLC.

Modbus and EtherNet/IP and DeviceNet

For EtherNet/IP (and DeviceNet), Modbus register and coil data has to be mapped to the Input and Output buffers that are cyclically transferred with the EtherNet/IP and DeviceNet Scanners. The register and coil data can be mapped to any part of the Input or Output buffers (assemblies in CIP terminology). Both the device mapping the Modbus data and the device receiving the data must understand the mapping.

You can find gateways from a number of vendors that can move data from Modbus Slaves or Modbus Servers and represent that data as EtherNet/IP or DeviceNet Adapter data. There are several RTA Gateways[3] that you can use to move data between a Modus TCP or Modbus RTU device and a Rockwell PLC.

---

[2] You can find these RTA Gateways at
http://www.rtaautomation.com/products/
[3] You can find these RTA Gateways at
http://www.rtaautomation.com/products/

MODBUS & OPC UA

It's hard to work almost anywhere in the automation industry without knowing something about OPC. It's been around forever and has been unarguably one of the most successful technologies to ever hit the factory floor.

OPC provides a standardized way to move data between two dissimilar systems as long as you are using Windows and Microsoft technology. OPC Servers embed the proprietary communications protocol of some automation device and make data in that device available to one or more OPC Clients. So, a Siemens PLC server on Windows embeds the proprietary S7 protocol and provides the data to OPC Clients like Historians, HMIs, Trend Analyzers and all sorts of other Windows devices.

DCOM (Distributed COM) is the basic Microsoft transport technology that provides the standard communications architecture between the Clients and Servers. DCOM allows Clients on one Windows platform to read and write in Servers on the same machine or on machines distributed across a network.

Tens of thousands of these Clients and Servers have been deployed for thousands of devices over the years. OPC has found its way into most factory floor architectures. In fact, if you know of an automation environment that doesn't use OPC in some way, I'd like to hear about it.

Well, OPC, which is now being called OPC Classic, is now being displaced by OPC UA. As a quick primer, here are the ten things that you should know about OPC Classic and the transition to OPC UA:

1. OPC Classic has one major deficiency, security—though even that deficiency is mostly mythical. It's really a lack of training and the lack of well-defined processes that frustrates users implementing OPC Servers.

2. OPC Classic relies on DCOM, and Microsoft obsoleted that technology a few years ago.

3. OPC UA is based on an SOA (Service Oriented Architecture) and the Web Services transport layer. The web services technology used in UA has a much more robust security layer than is found in DCOM.

4. OPC UA combines the Data Access, Historian, Alarm and Events, and all the rest of the OPC Classic specifications into a single entity with an Object-based model. This architecture is a huge step forward over Classic OPC.

5. The biggest advantage to having OPC UA in a device is the ability to provide seamless communication to systems based on Linux and other non-Microsoft platforms. This is very important as more and more business systems require communications with factory floor devices, and many of those business systems are some flavor of Linux.

6. OPC UA Client software devices will probably include mechanisms to ease the transition from Classic to UA.

7. OPC was not incompatibility-free. There were always incompatibilities between different specifications, different

specification versions, and more. And unfortunately, when you start introducing OPC UA, you are going to find still more incompatibilities to deal with. One of the products that you'll probably need is a UA to Classic translator.

8. Siemens is a big fan of UA. They strongly believe that OPC UA Servers in their PLCS will provide the network architecture of the future; a network architecture with seamless communication between those PLCs and Automation and Business systems.

9. OPC UA is probably going to reduce your costs and hurt the business of OPC Class Server providers like Matrikon, Kepware and others. You can expect that PLCs and many other automation devices like drives, motion controllers and other advanced devices will probably come with OPC connectivity as a standard option.

10. Security is a growing concern. There are threats from everywhere. Threats from high school kids who want to mess with your packaging machine to state-sponsored terrorists who want to take over the power grid, the water system of your city, or the FAA flight management system. UA provides much better security than classic, so many think that UA is going to be deployed just because of this.

The biggest reason for UA to win at the factory floor to IoT or Enterprise communication system is the infrastructure of companies and trade associations that are supporting it: Oil and gas, BACnet, PLC Open and the EtherCAT consortium; vendors like Siemens; huge IT companies like SAP. With all this momentum, it looks like the protocol of the future.

So how does an old guy like Modbus RTU or its younger brother, Modbus TCP, fit into this new architecture? Is Modbus relegated to the trash heap of history? Is it no longer valuable?

The answer is unequivocally NO. Modbus is alive and

well and will be around for quite a bit longer. There are so many thousands of products that have Modbus RTU or Modbus TCP interfaces that it will quite literally take fifty years before they're all obsoleted.

What will happen is that gateway devices will be made available to grab Modbus register data from Masters, Slaves, Clients and Servers and make that data available in OPC UA Servers where it can be grabbed. Grabbed by Enterprise systems, database systems, productions systems, maintenance systems…anyone who might need to access that Modbus Register data.

What you will find is that OPC UA gateways won't just map a single register to a 16-bit unsigned value in the OPC UA server. No, what will happen is that the Register and Coil data will be combined, manipulated, and metadata a will be added so that:

On Modbus: a 16-bit unsigned Temperature with a value of 1609 will be mapped to an OPC UA Object with the following structure:

```
Object Name: Cooler Temperature
Data Variable Value: 2.5316
Data Variable Type: Floating Point
Range Property: 0 to 30
Scaling Property: 5:1
Type Property: Celsius
```

Notice that much more data will be available from the OPC UA node data than was available in the Modbus device controller. This will require vendors or users to map their data more precisely, but it will allow systems farther up the food chain to process data more efficiently and correctly than if they just had a raw integer value in the Modbus register.

And that will make the data more useful and more valuable. And that's the power of OPC UA.

MODBUS WILL LIVE FOREVER

When I think of Modbus and the future, I am reminded of that old joke that everyone wants to go to Heaven but no one wants to die. I am reminded of that joke because I often think that Modbus is obsolete and won't last, but then I realize that Modbus will never die.

Yes, the Internet and Web Services are dominating our thinking and our new development. Everything has to be connected to the web. Everything has to connect to the Enterprise. Every miniscule bit of data must be manipulated, analyzed and archived.

I won't argue that this isn't the direction of the future, but I don't think that means that every photo eye, proximity switch and valve is going to be web-enabled. There are some things that are too low in the automation hierarchy to be of consequence. Who wants to manage Enterprise communication and the bandwidth needed to move data from a thousand small devices in a manufacturing plant? How about a million devices in refinery?

The traditional automation hierarchy will change, but that is going to happen fairly slowly. The Programmable

Controller is going to evolve, but it will evolve slowly. Until it does, we will have an automation hierarchy that is substantially similar to what we have today. DeviceNet, PROFIBUS DP, BACnet, EtherNet/IP, PROFINET IO and the other I/O protocols are not going away. Just as the networks that feed IO to Programmable Controllers like Io-Link aren't going away.

There is always going to be a place in this automation structure for small, single-purpose devices that support a protocol like Modbus RTU or Modbus TCP. The advantages of using Modbus technology haven't changed, and it's still going to be the right choice for years to come:

1. It's small. You can easily fit it into the smallest of processors. No big RAM or ROM requirements.

2. It's cheap. You can build it yourself or you can buy it for a very small charge.

3. It works with all processors. Anything that has a UART can run Modbus.

4. It's simple to deploy. It's just daisy-chained RS485. Wire from one device to the next to the next or a standard Ethernet network.

5. It has very simple data typing, memory map and operating functions so that most competent programmers can implement it within a week or so.

6. There's a lot of it out there! You can't go wrong with a Modbus interface. Everybody understands it and can usually find a way to move the data to where it needs to go.

When I think about this, I come to the conclusion that MODBUS WILL LIVE FOREVER. A big reason for that is the guys that build a lot of automation devices are not network communication experts. They are domain experts in valves, glue controllers, pumps, chillers, drives or whatever. They need a way to talk to their device. Something simple that isn't going to cost too much and that they can get implemented quickly. That answer for the last 40 years has been Modbus, and I see no reason for it to change now.

Hail Modbus! Modbus Forever!

MODBUS & REAL TIME AUTOMATION

Real Time Automation has been called quirky, and I think that's a pretty good description. I'm certain we're the only automation company where you get a Cracker Jacks-type prize in the box with our products. I'm certain we're the only company with a newsletter where the owner writes about health, friendship, gratitude and getting swindled on the streets of Italy. I'll bet that we're the only automation company you'll find that when you call for support you often find yourself talking to one of the product developers. It's a company where new interns are introduced to "KAN JAM" on their first day on the job. I could go on, but that gives you an idea what "quirky" means.

But the big question is "Why do our customers buy from us?" And continue to buy more every year. Well, there are three reasons for that. First, there is ease of use. Every company says that their products are "easy to use," but I'm telling you that doesn't make it easy to use. What makes a product easy to use is a lot less about technology and more about people. Specifically, we don't let our engineers engineer the user interface. And that means a lot. An engineer will always focus more on the product development than on the customer. We have a "High Tech Anthropologist" that makes sure the code does what the customer needs it to do, not what the code needs the customer to do.

The second reason our customers order from us is that we provide products that meet their needs. Most companies design products that make distribution easy: Put everything in one box and let the customer figure it out. Then they only need to ship a single box to everyone. We ship a box that does exactly what the customer wants. If the customer wants a product that gets them a specific result, we send them a product that is as pre-configured as possible to do that. We don't send one that does 500 things and the customer has to figure out how to disable or avoid the other 499.

The third reason is our support. We're located in the US central time zone and available during normal working hours for the entire country. I've been on the factory floor trying to get a machine running. I know the frustration, panic and disappointment when you can't get the help you need to get that machine running. I've promised myself that we're never going to be one of those companies. You can count on us, and if we've ever disappointed you on that, I want to know about it.

So what problems do we solve for customers? Our best selling products, at the time of writing of this book, are the gateways that we offer. We provide gateways for all sorts of Industrial Automation technologies, including:

- ✓ EtherNet/IP
- ✓ PROFINET IO
- ✓ DeviceNet
- ✓ OPC UA
- ✓ N2

✓ Modbus TCP

✓ Modbus RTU

✓ …and others

Since this is a Modbus book, let's look at some of the solutions you can get from Real Time Automation (www.rtaautomation.com/products) to solve your Modbus connectivity problems:

Gateways for Modbus RTU Slaves
Connect Modbus RTU Slaves to a Modbus TCP/IP Client
Connect Modbus RTU Slaves to a BACnet/IP Client
Connect Modbus RTU Slaves to a DeviceNet Master
Connect Modbus RTU Slaves to an Allen-Bradley PLC
Connect a Modbus RTU Slave to an ASCII Device
Connect a Modbus RTU Slave to a TCP/IP Device
Connect a Modbus RTU Slaves to BACnet MS/TP
Connect a Modbus RTU Slave to an Allen-Bradley PLC
Connect a Modbus RTU Slave to Connect a Modbus TCP/IP Server
Connect a Modbus RTU Slave to EtherNet/IP Scanner

Gateways for Modbus RTU Masters
Connect DeviceNet Slaves to a Modbus RTU Master
Connect a Modbus RTU Master to an EtherNet/IP Scanner
Connect a Modbus TCP Servers to a Modbus RTU Master
Connect a Modbus RTU Master to a BACnet/IP Client
Connect a Modbus RTU Master to a Connect a Modbus TCP/IP Client
Connect a Modbus RTU Master to ASCII
Connect a Modbus RTU Master to EtherNet/IP Adapter
Connect a Modbus RTU Master to DeviceNet Master
Connect a Modbus RTU Master to Ethernet TCP/IP
Connect a Modbus RTU Master to PLC

MODBUS TCP/IP Server Gateways
Connect a Modbus TCP/IP Server to BACnet/IP Client
Connect a Modbus TCP/IP Server to EtherNet/IP Scanner
Connect a Modbus TCP/IP Server to Connect a Modbus RTU Master
Connect a Modbus TCP/IP Server to ASCII
Connect a Modbus TCP/IP Server to ASCII Four Port

Connect a Modbus TCP/IP Server to Allen-Bradley PLC
Connect a Modbus TCP/IP Server to DeviceNet Master
Connect a Modbus TCP/IP Server to BACnet MS/TP Master
Connect a Modbus TCP/IP Server to Connect a Modbus RTU Slave
Connect a Modbus TCP/IP Server to PROFINET IO Controller

MODBUS TCP/IP Client Gateways
Connect a Modbus TCP/IP Client to ASCII
Connect a Modbus TCP/IP Client to ASCII Four Port
Connect a Modbus TCP/IP Client to Connect a Modbus RTU Slave
Connect a Modbus TCP/IP Client to DeviceNet Slave
Connect a Modbus TCP/IP Client to Connect a Modbus RTU Master
Connect a Modbus TCP/IP Client to BACnet/IP Client
Connect a Modbus TCP/IP Client to EtherNet/IP Adapter
Connect a Modbus TCP/IP Client to Ethernet TCP/IP
Connect a Modbus TCP/IP Client to DeviceNet Master
Connect a Modbus TCP/IP Client to Allen-Bradley PLC
Connect a Modbus TCP/IP Client to EtherNet/IP Scanner
Connect a Modbus TCP/IP Client to PROFINET IO Controller

For more information on these products, visit
http://www.rtaautomation.com/.

You get three specific promises when you select one of these products from Real Time Automation.

One, we'll answer the phone with a live person who'll send you right to the person you need.

Two, we have what you need in stock and we'll do everything we can to get it out today if you need it first thing in the morning. Most days we ship at 3pm central, but often we'll be taking boxes to UPS at 5:30pm or 6. And once we drove 8 hours to get it there by midnight!

My third promise is that you will receive a product that you can easily use. That means that the UI is understandable, and configuration is more like a dream than a nightmare. We are exceptionally vigilant about this point because we understand that the cost of our product is nearly nothing compared to your time to install, commission and put in into operation. We're not perfect, but we always work toward it.

Don't forget about the wealth of resources waiting for you on the resources web page for this book. Just visit either of the following web pages:

http://www.rtaautomation.com/technologies/modbus-rtu/
http://www.rtaautomation.com/technologies/modbus-tcpip/

ABOUT THE AUTHOR

John Rinaldi, is Chief Strategist, Business Development Manager and CEO of Real Time Automation (RTA) in Brookfield, WI.

After escaping from Marquette University with a degree in Electrical Engineering (graduating cum laude, no less), John worked in various jobs in the Automation Industry before once again fleeing back into the comfortable halls of academia. At the University of Connecticut, he once again talked his way into a degree, this time in Computer Science (MS CS).

John achieved marginal success as a Control Engineer, a Software Developer and IT Manager before founding Real Time Automation because "long term employment prospects are somewhat bleak for loose cannons."

With a strong desire to avoid work, responsibility and decision making, John had to build a great team at Real Time Automation. And he did. RTA now supplies network converters for industrial and building automation applications all over the world. With a focus on simplicity, US support, fast service, expert consulting, and tailoring for specific customer applications, RTA has become a leading supplier of gateways worldwide.

John freely admits that the success of RTA is solely attributed to the incredible staff that like working for an odd, quirky company with a single focus: "Create solutions so simple to use that the hardest part of their integration is opening the box."

John is a recognized expert in industrial networks and the author of three books, two on Industrial Networking. The Industrial Ethernet Book focuses on explaining Industrial Ethernet concepts in a straightforward, clear fashion. The same simplicity is found in John's other book, OPC UA: The Basics, an overview of the enhancements to OPC technology that allow for Enterprise communication.

You can reach John here:

John Rinaldi
Real Time Automation
N26W23315 Paul Rd. Pewaukee, WI 53072
Brookfield WI 53005
262-436-9299(office)
414-460-6556 (cell)

http://www.rtaautomation.com/contact-us/
https://www.linkedin.com/in/johnsrinaldi

JOHN RINALDI

www.ingramcontent.com/pod-product-compliance
Lightning Source LLC
Chambersburg PA
CBHW070910180526
45168CB00005B/1996